The Do-It-Yourself GUNPOWDER COOKBOOK

Don McLean

PALADIN PRESS • BOULDER, COLORADO

The Do-It-Yourself Gunpowder Cookbook
by Don McLean

ISBN 13: 978-0-87364-675-8
Printed in the United States of America

Published by Paladin Press, a division of
Paladin Enterprises, Inc.
Gunbarrel Tech Center
7077 Winchester Circle
Boulder, Colorado 80301 USA
+1.303.443.7250

Direct inquiries and/or orders to the above address.

CONTENTS

PART ONE
GROW YOUR OWN GUNPOWDER
1

Chapter 1
FROM WHENCE THIS DEMON CAME
5

Chapter 2
THE WHAT, WHY, AND HOW-TO OF BLACK POWDER
13

Chapter 3
"RED-OR-WHITE" PROPELLANT POWDER:
HOW SWEET IT IS
19

PART TWO
HOME-CANNED CHEMISTRY
FOR SURVIVAL POWDER MAKING
23

Chapter 4
SALTPETER
27

Chapter 5
CHARCOAL
37

Chapter 6
SULFUR
43

Chapter 7
SWEET SOLUTIONS
57

Chapter 8
THE MINIMALIST MINER
61

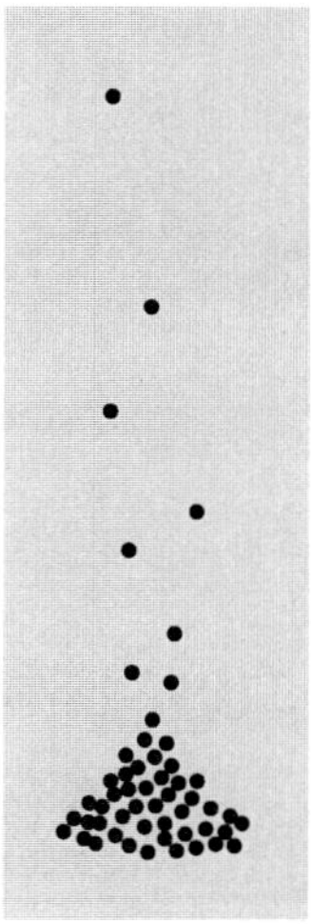

WARNING

Whenever dealing with gunpowder, explosives, chemicals, and specialized equipment, machinery, and procedures, special precautions *must* be followed in accordance with industry standards for experimentation and production. Failure to strictly follow such industry standards may result in harm to life and limb.

Therefore, the author and publisher disclaim any liability for any damage or injuries of any type that a reader or user of information contained in this manual may encounter from the use of said information. Use this manual and any end product or by-product at your own risk. This manual is *for information purposes only.*

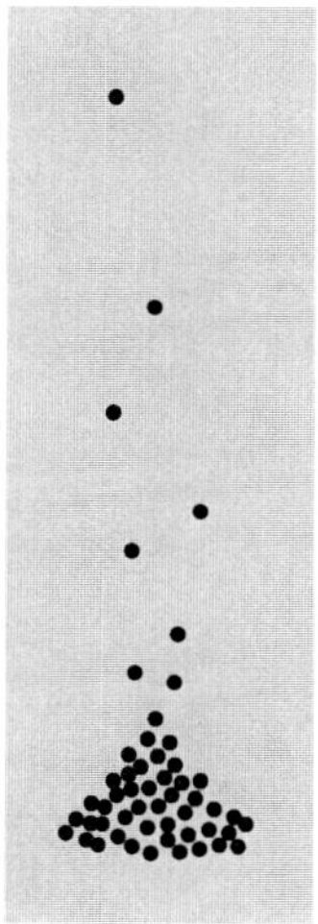

PART ONE

GROW YOUR OWN

GUNPOWDER

Faced with a survival situation, a man or a nation often reverts to lower levels of technology as available technology—which is dependent upon support from an intact infrastructure—falls away. But there is no need to regress to aboriginal levels if one has first stockpiled knowledge. Very basic hand tools and a toolbox of knowledge will enable one to maintain at least a level that is roughly equivalent to the early Industrial Age. And that's just fine with a lot of us curmudgeons, who probably would have been a lot more comfortable in that much simpler era anyway.

In those days, materials were wrest from the earth with hand tools and animals and simply processed with other hand tools and possibly more animals. The final product was simple, basic, and useful. It was within the reach of anyone who knew what he was doing, not only those blessed with a technological infrastructure.

A case in point is black gunpowder, the oldest propellant/explosive known. It is still in use today because it is still useful today—in certain applications so uniquely useful, no substitute has been found for it. And the chemical properties of black gunpowder have not changed since it was discovered, which means that if you must, you can make it the same way it has been made for hundreds of years. With a small assist from modern processing methods, however, the same result can be had at considerably less risk and with less machinery than was required then.

The purpose of this monograph is to outline the methodology to win the materials from the earth itself and produce a useful propellant/explosive powder using the simplest of tools. This was done hundreds of years ago, and you can do it today. And with modern compounding methods, you can do it more easily and more safely.

Chapter 1

FROM WHENCE THIS DEMON CAME

The oldest explosive known is black gunpowder. It is impossible to determine with exactitude the *real* date gunpowder was invented, or when it was first used in a firearm to throw a projectile. There has been an interminable debate over who *really* invented it or, if it had more than one inventor, who invented it first. Ever since it appeared on the scene in feudal Europe and dramatically changed the face of same, the schools of thought generally have followed support of a local hero. A careful review of literature, ancient and modern, reveals with clarity only the necrophagous tendencies of those who write on the subject, and the only certainty is that, whatever the date of black powder's invention, xenophobia was invented first.

It is this writer's opinion that this particular wheel was probably reinvented a number of times by various experimenters, and that the development of firearms as such was not predicated upon the availability of black powder or the components for its formulation as much as it was on the availability of metallurgy and fabrication techniques for suitable weapons to shoot it in. Black powder was probably first used in the Orient—India or China—as an explosive and rocket propellant, but it was probably first used in Europe in what we would today call a firearm.

The following are the most favored accounts. You may believe the one you like, or if you prefer, you may do like everybody else and make up your own version to best suit your particular revisionist view of history.

SCHWARZ, BACON, AND ETCETERA, INC.

Early German and Central European literature will assure you that black powder was accidentally invented/discovered by Berthold Schwarz, O.S.F., a Franciscan monk in Freiburg. According to this rendition, Schwarz, a passionate alchemist, was one day pounding together a mixture of saltpeter, sulfur, and charcoal in a mortar. Suddenly the mixture exploded, blowing the top off the mortar and flinging the unfortunate monk to the ground. It is generally accepted that Berthold Schwarz was the first to openly publish the results of his experimentation with black powder, and once he had done that, *everybody* wanted to play with it. The resulting development and use of cannon changed the face of feudal Europe forever.

English literature, and the literature of those who believe what the English write, will assure you that black powder was accidentally invented/discovered by Roger Bacon, a Franciscan monk at Oxford. According to this rendition, Bacon, a passionate alchemist, was one day pounding a mixture of saltpeter, sulfur, and charcoal together in a mortar. Suddenly the mixture exploded, blowing the top off the mortar and flinging the unfortunate monk to the ground. And then the plot thickens . . . .

Bacon reportedly was of the opinion he had unleashed powers too great to be entrusted to the great unwashed, so he wrote the results of his experiment in an obscure Latin code. Unfortunately, these notes (the Baconists will assure you) were uncovered by Berthold Schwarz (the rat), who broke the code and stole the idea, then had the gall to publish the results as his own. Bacon's formula called for an intimate mixture (by weight) of 41.2 percent saltpeter, 29.4 percent sulfur, and 29.4 percent charcoal, a fairly weak mixture when compared to the stoichiometric proportions of 74.64 percent saltpeter, 13.51 percent charcoal, and 11.85 percent sulfur.

Both the above accounts are probably apocryphal, however, since *unconfined* black powder does not explode but only produces a violent conflagration. Even Baconist texts express the doubt that Bacon had a handle on how useful this concoction would be as a propellant, but the idea and the formula as published by Schwarz—his or not—provided the springboard for other tinkerers, and within mere decades the nearly universal and still current proportions had been empirically worked out to 75 percent saltpeter, 15 percent charcoal, and 10 percent sulfur, by weight, strikingly close to the theoretical stoichiometries above derived from atomic weights.

One historical point that has always given me pause is, why one fine day did Schwarz and/or Bacon happen to find themselves pounding saltpeter, charcoal, and sulfur together in a mortar? It seems probable that this was a further development of the "Greek Fire" incendiary mixture that the Byzantines made from Indian Saltpeter (aka "niter" or potassium nitrate, KNO_3), sulfur (aka "brimstone," S), and crude petroleum (aka "naphtha"). This mixture could not be extinguished (because it carried its own oxygen in the saltpeter) and would burn on or under water. It was used to good effect against the invading Saracens.

ANCIENT ORIENTALS, INC.

Explosive mixtures are known to have been used by the ancient Chinese and East Indian civilizations fairly early in their history, and a hundred years ago opinion prevailed in the West that the Chinese had been the first to invent gunpowder and the first to use it in firearms. One Chinese manuscript makes reference to weapons used in the siege of Pienkingu in the thirteenth century B.C. It describes how these weapons, called *huo-pao*, fired projectiles that made a terrible noise and set the fortifications of the town on fire. It is also assumed by some that the *to-lo-siang*, or "lance of indomitable fire," was the first gun, and the *tsin-tien-lei* the first explosive shell.

As described, these devices could have been either rockets, which it is generally acknowledged the Chinese had, or cannon, although the state of metallurgy in China at that time would not have lent itself to building cannon capable of throwing explosive shells great distances. Supporters of the "powder and cannon came from China" thought conclude that the secret of gunpowder and its uses was most probably brought to Europe by the Mongols, although no explanation is made as to why only the powder and not the cannon came into immediate use there.

Another theory popular with historians of the last century was that gunpowder was invented even earlier in India, and some nineteenth-century historians went so far as to claim that Alexander the Great himself had been familiar with its effects.

Yet another hypothesis was that gunpowder was introduced into Spain by the Moors, and by the Byzantine Greeks into other European countries. In support, it is pointed out that gunpowder was used in Spain and Greece much earlier than in the countries of central and northern Europe.

TAKE YOUR PICK

In examining all these worthy lines of thought, one should bear in mind what is known of military history and the evolution of military techniques. Even in their crude early (and largely ineffective) forms, cannon created quite a splash among contemporary historians for the terror they caused. Because they were important events, ancient battles were well recorded, and it would seem improbable that cannon or other firearms would have escaped the note of contemporary writers, had such weapons existed. Ancient references in Oriental texts refer to what *might* be construed as cannon, but more plausibly would have been rockets and crude bombs, and it is worthy to note that the Orient's development of small arms lagged well behind Europe once gunpowder was made in the West. Crude matchlocks were still in vogue in China and Japan well into the nineteenth century, long after European gun makers had progressed from the matchlock to the wheel lock to the flintlock and were well on the way to developing percussion arms.

Using the wealth of military history as a sounding board (but keeping in mind it was written by the victors) leads this writer to conclude that in all probability the nations of the Far East first discovered gunpowder, then developed explosives, rockets, and crude bombs, but that the firearm as we would recognize it was developed in Europe. It is interesting to note that until the time of Bacon and Schwarz, saltpeter was not available in Europe except as a commodity that was purchased from the Orient, and to this day potassium nitrate is called "Indian Saltpeter."

EVERYBODY'S MAKING IT

Once the techniques for making black powder and weapons suitable to its use became widespread in the latter 1200s, techniques tended to become universal both through the exchange or leakage of information and because everybody tended to do what worked. Early techniques for making black powder were pretty much the same as described by Bacon, Schwarz, et al. The three components were pounded together in a mortar or horse/water/steam-powered "stamp mill," where the pestles of hardwood mortars were raised and tripped by machinery. Pounding an explosive to thoroughly mix it is a dangerous procedure, and stamp mills were outlawed in England in 1722 except for use in the production of fine sporting powders.

The first mixtures were indeed true powders, and the loading of a cannon was tricky business because the powder had to be packed just right: too firm and it burned too slowly; too loose and the conflagration spread so fast that pressures rose too quickly and the cannon would burst before the ball had a chance to move. This problem was first addressed by using a small ball in a big barrel, the extra "windage" allowing a margin of safety. But this made for inaccurate, short-range guns. The problem of controlling the rate of burning was whipped early in the fifteenth century when the process of "corning" (forming a solid cake, then breaking the powder into grains) was developed.

In this process, the mixture was moistened while being mixed in the mortar, either with alcohol or urine. Some, perhaps hoping for the best of both, specified the urine of a wine drinker be used. The paste was then pressed into a cake, dried, broken into chunks by hand, and then further broken and screened by metal or hardwood balls over a screen. Because of the varying proportion of surface area, the various screen sizes had varying burning rates and were used for various purposes. Cannon powder was the largest, priming powder the smallest.

The final step in slow, even-burning cannon powder was General Rodman's (inventor and U.S. Army ordnance technician) development in the late 1800s of "cocoa" powder, which used undercharred rye straw for charcoal, with a ratio of 80 percent saltpeter, 16 percent undercharred rye straw, and 4 percent sulfur, with the grains formed by pressure into hollow hexagonal prisms. Such slower powder formulations are still useful for slow-burning powders designed to heave instead of shatter, as for mining gemstone.

Other evolutionary developments included the introduction of wheel mills to replace the unsafe stamp mills, other methods of pressing the powder into complex geometric shapes to regulate surface area/burning rate, and coating the powder with graphite to keep it from absorbing water. Also, all possible sources of ignition or spark creation in the workplace were eliminated one by one as powder factories blew up one by one.

Sparks so small as to be unnoticed are all around us in everyday modern life, and not only does the survivalist-cum-powdermaker have to contend with all the ignition sources that have killed uncounted legions of powder makers over the past eight hundred years, but with the modern ones as well. The old-timers didn't have to contend with such things as static-making synthetic clothes, including skivvies and socks, and induced sparks from radio transmissions, and they still had a short life expectancy. They soon learned to keep the powder wet through as many steps of the manufacturing process as possible, only use soft (usually copper or zinc) metal around the plant, keep the batches small, never store a lot of it together, and when all else failed, as they knew it would, build factories and magazines with earthen walls around them and flimsy roofs so the blast would not spread to adjacent mills and magazines.

SAFETY FIRST, LAST, AND ALWAYS

The technique for avoiding fires and/or explosions is to preclude, by proper planning and procedures, every possible ignition source. But the only real way of ensuring your survival when making black powder is to ensure you will survive the accident, because if you keep at it long enough, sooner or later you will have one.

First, remember that black powder burns, it does not detonate. Small quantities will not explode unless confined in a strong container that allows it to build up pressure, and a fire is preferable to an explosion. So **Rule Number 1** is: Never put or store black powder in a strong container. Use airtight, soft plastic vitamin bottles, or the very small coffee cans with plastic lids.

Rule Number 2: Never store a lot of it in the same place, or inside a dwelling, or near flammables, or in a metal container if it might be subject to induced currents, even from radio transmissions.

Rule Number 3: Always assume that, sooner or later, something will set it off, so work away from any dwelling or flammables, keep quantities small enough to survive, and wear a welder's apron, safety glasses or face shield, and gloves so you won't get burned too badly.

Rule Number 4: Eliminate one by one every possible source of ignition from your workplace. No hard-metal tools. No wool or synthetic fibers in clothing, brushes, brooms, or carpet. No flames or smoking. No radio transmitters. No open electric motors.

Rule Number 5: Read Rule Number 3 again, and believe it.

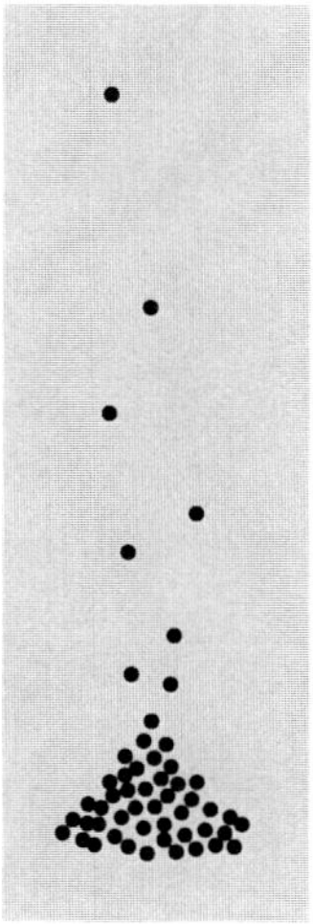

Chapter 2

THE WHAT, WHY, AND HOW-TO OF BLACK POWDER

Black powder is classified as a low explosive—that is, it is more easily ignited but is not as powerful as "high" explosives. It is an intimate mechanical mixture of fuel (charcoal and sulfur) and an oxidizer (potassium or sodium nitrate) that will burn quickly and independently of its surroundings to produce copious amounts of hot gas. If confined, as in a gun barrel or a borehole in rock, these products of combustion will cause an explosion. Black powder does not of itself detonate unconfined as do nitroglycerin, TNT, PETN, or "high" explosives. The burning rates of black powder are varied by grain size (the smaller, the faster), grain density (the more dense, the faster), and by adjusting the ratio of fuels to oxidizer.

Because the oxidizer must be very thoroughly and intimately mixed with the fuels, the old-timers mixed it for a long time, increasing the likelihood of its ignition, particularly in a dry stamp mill. Later on, the practice of keeping the mix damp both reduced the danger of ignition *and* tended to impregnate the porous charcoal with saltpeter solution.

The benefits of this damp-mixing technique (safety and intimate mixing) were picked up and perfected several years ago by technicians developing field procedures for behind-the-lines black-bag artists, who would have need to make their own black powder from materials at hand. The procedure that was developed entails mixing the ingredients in a water medium, which ensures complete saturation of the fuels with the oxidizer, then "salting out" the desirable mixture by introducing it into an organic solvent. This sophisticated but very simple technique is eminently suited to the production of black powder in a survival situation, as it requires only everyday equipment, produces a predictable and uniform product, and is about as safe as it gets. Which is to say, of course, not very safe at all if you don't strictly adhere to all safety procedures.

Because the premise was that a field operative would not have access to equipment such as accurate scales, the usual stoichiometric weights were converted to volumes. The volumes given below are in cups and quarts, but any system of volumetric measure can be used as long as the proportions are kept constant. An additional benefit is that the saltpeter need not be finely ground for this technique.

THE FRANKFORD ARSENAL FUDGE FACTORY RECIPE FOR BLACK POWDER

This specific recipe for field-expedient (read that *survival situation*) black powder will produce nearly 2 pounds of powder that is suitable for use as blasting or musket powder.

The supplies and materials you will need are:

- a heat source (an electric hot plate is preferred for safety reasons, but any heat source will do)

- a 2-gallon plastic or metal bucket
- a 4-quart skillet or saucepan
- a 1-foot-square section of window screen on a wood frame
- a large wooden spoon or flat stick, such as a piece of shingle
- a plain-weave cloth at least 2 feet square, such as a piece of bed sheet
- an 8-ounce measuring cup
- 3 cups of granulated or powdered potassium nitrate ("Indian Saltpeter," KNO_3); sodium nitrate ("Chilean Saltpeter," $NaNO_3$) can be substituted if necessary
- 2 cups of finely powdered hardwood charcoal (dogwood, willow, alder), settled down by tapping the cup, but not tamped or compacted
- 1/2 cup of finely powdered sulfur
- 3 cups of water
- 10 cups (2 1/2 quarts) of 70-percent isopropyl alcohol; wood (methyl) alcohol will substitute, as will vodka if that's all you have

The compounding procedure is simpler than baking a cake, but must be followed *exactly*:

1. By volume, measure 3 cups granulated potassium nitrate, 2 cups powdered charcoal, and 1/2 cup of powdered sulfur into the 4-quart pan or skillet. Moisten with 1 cup of water. Using a wooden stick or spoon, thoroughly blend the ingredients. Add an additional 2 cups of water and stir thoroughly.

2. Place the pan containing the above slurry on the heat source and stir gently to maintain a homogeneous mixture until the first bubbles of a simmer begin to form. *Do not boil the mixture, and do not allow any mixture to dry on the side*

of the pan, as any dry powder might be ignited by the heat source.

3. While stirring the receiving bucket vigorously, quickly pour this mixture into 10 cups of alcohol that have been previously placed in the 2-gallon bucket. Stir this mixture briefly, then allow to stand for 5 minutes.

4. To collect the black powder, strain the entire contents of the alcohol bucket through the cloth. Remove as much liquid as possible by wrapping the cloth around the powder, twisting it up, and squeezing the resulting bag, as you would do to collect juice from fruit pulp when making jelly.

5. Spread this wet powder on a flat surface, smooth it out evenly to 1/2 inch thick, and allow to dry to a damp solid. It should be moist enough that a handful squeezed tightly together will form a coherent lump in the hand, but no moisture can be brought to the surface by squeezing and the powder will not stick to the hand.

6. While the above is drying, wash and dry the bucket and place the screen over it. When the moist powder from step 5 is properly dried, place a handful on the screen and gently granulate it by rubbing the damp solid through the screen by hand. If the particles collected in the bucket appear to stick together or change in shape (i.e., slump flat or form into puddles), the mass was too damp. If this happens, recombine the entire batch, redry as in step 5, and repeat the granulation operation.

Dry this granulated black powder completely by spreading it gently on a flat surface in a layer no more than 1/2 inch deep. Sun drying is preferred in the field because the black powder, being black, will be heated to warmer-than-ambient temperatures by the sun, thus ensuring lower-than-ambient moisture content. It may also be dried in small batches in a fruit dryer (*not in the house*) at about 150 degrees Fahrenheit for three hours. Once dried, it must be kept in airtight containers. The more quickly the powder is dried, the better quality it will be, but *never* dry it over a direct heat source, and keep the batches small and unconfined.

NOTE: Save the solution left over from step 4. It can be evaporated to dryness to reclaim any saltpeter. *Bear in mind that, depending on the fineness of the filter cloth you used, there may be microparticles of charcoal and sulfur in this dried saltpeter residue, and when dried it should be handled as an explosive mixture.*

When made with technical-grade components using the procedure above, this nominal 10-mesh powder will have a burn rate of approximately 10 to 12 centimeters per second. It must be stored, handled, and used following all the established safety procedures for black powder. If you have never used black powder, do *not* whip up a batch and play with it to learn what it's like. Study up on the subject, learn with commercial powder, and save "survival" powder making for "survival" situations. Storage and transportation of black powder in quantities over a certain amount are subject to Bureau of Alcohol, Tobacco and Firearms (BATF) and Interstate Commerce Commission (ICC) regulations.

For deep-hole blasting or cannon powder, a coarser grain size is made by substituting a coarser screen. For finer grain sizes, such as for priming a flintlock, a tighter screen may be used, or the powder very carefully ground (in small batches!) in a wooden mortar and then screened. Grind with a gentle rotary motion. *Never pound dry black powder, even in an open mortar.*

Chapter 3

"RED-OR-WHITE" PROPELLANT

POWDER: HOW SWEET IT IS

As you will see in the following chapters, extracting saltpeter from natural soils takes little more than buckets, shovels, and the right technique. Preparing charcoal is also relatively simple. But unless you live next to a sulfur pit, extracting sulfur from minerals, although it is perfectly doable when you know how, can be "labor intensive." So it's good to have some alternatives up your sleeve.

"Red-or-white" powder is just such an option, as it does not require any sulfur, just saltpeter, sugar, and ferric oxide (red iron rust, Fe_2O_3) if you have it. In a from-the-ground-up survival situation, all the components can be made from natural sources: saltpeter from soil, sugar from maple sap or sugar beets, ferric oxide from any rusted iron. This time, we will assume you have saltpeter, sugar, and iron oxide on the shelf.

Red-or-white powder is a suitable propellant for rockets

and small arms of less than .50 caliber. For reasons unknown to this writer, the technicians in the Skunk Works at Frankford Arsenal did not deem it suitable for pistol use or as a blasting agent. The reason may be that the finished product is quite hygroscopic.

THE FRANKFORD ARSENAL FUDGE FACTORY FUDGE RECIPE

The materials you will need for red-or-white powder are:

- a heat source suitable for cooking
- a 2-gallon metal bucket
- an 8-ounce measuring cup
- wood spoon and/or rubber spatula
- metal (cookie) sheet or aluminum foil
- at least 1 square foot of window screen
- 2 1/3 cups of granulated potassium nitrate (saltpeter)
- 2 cups of granulated white sugar
- 1/8 cup of powdered ferric oxide (if available)
- 3 1/2 cups of potable water

The compounding procedure for this propellant, like the ingredients, is closely akin to making fudge:

1. Place the water in the bucket over a low heat and add the sugar, then the saltpeter. Stir occasionally until the sugar and saltpeter dissolve completely.

2. If available, add the ferric oxide (rust). Increase the heat until the mixture boils gently; stir occasionally to prevent scorching. If you added ferric oxide, the mixture will have a rusty orange color.

3. Stir and scrape the sides and bottom of the bucket occasionally until the mixture is reduced to 25 percent of its original volume, then stir continuously.

4. As the mixture boils down, it will become thicker and

thicker until it is the consistency of homemade fudge. When it is about like breakfast porridge, remove the mass from the heat and spread it on the metal sheet, no more than 3/8 inch thick.

5. While the material cools, score it in a crosshatch grid pattern in 1-inch or less squares to aid drying.

6. Allow the material to dry in the sun or on a very low heat source. As it dries, rescore it every 20 minutes or so in a new pattern to expose fresh material.

7. When the material has dried to the point where it is soft and moist but not sticky to the touch, it can be granulated. Place a small spoonful on the screen as you did with the black powder and rub it back and forth with a spoon or other flat object. The material will come through the screen as small wormlike granules and be caught in the bucket.

8. Once granulated, the material is respread very gently on the metal sheet and air-dried *completely*—that is to say *bone-dry*. It is very hygroscopic and must be protected from damp air or any other source of moisture.

If allowed to absorb moisture, this propellant delivers erratic results, and if moist to the point of becoming softened it will not load properly. If it becomes moist to the point of stickiness, it must be regranulated and redried. Coating the granulated powder with powdered graphite helps to keep out the moisture somewhat, but if this is done then it cannot be regranulated satisfactorily. In short, it adds a new dimension to the old adage, "keep your powder dry."

NOTE: The red-or-white powder described above should *not* be confused with the "white gunpowders" that use potassium chlorate as an oxidizer, which first made their debut in the late 1800s. These were found to be too unstable to be usable then, and they still are.

PART TWO

HOME-CANNED CHEMISTRY

FOR SURVIVAL POWDER MAKING

We have discussed techniques for making your own gunpowder via methods that offer the best chances for safety. Now we will outline how to make the components for gunpowder from common materials likely to be found on your front lawn. If you have a willow tree, manure, and a plaster-of-paris flamingo, you're all set. Or, if you are patient, you can use dead cats, recycled beer, and your living room ceiling.

Since the early makers of black powder did so without access to a chemical supply house, a person in a survival situation can do the same by copying the old techniques, but adding such enlightened procedures as have been developed by the science of chemistry over the past eight hundred years. Fortunately, black powder is a simple formulation of only three components, and we will deal with them one at a time.

Chapter 4

SALTPETER

Both Indian Saltpeter (potassium nitrate) and Chilean Saltpeter (sodium nitrate) can be used as the oxidizer in black powder. Because of its somewhat greater propensity to absorb water from the air, however, sodium nitrate is decidedly a second choice. In fact, when sodium nitrate deposits were first discovered in South America, the usual application was to react it with potassium chloride to make potassium nitrate. Sodium nitrate simply was found unsuited for military purposes or for any black powder that would be stored for a long time. Sodium nitrate is so hygroscopic it can absorb enough water from the air to dissolve into a puddle of its own making.

Although originally procured from the Orient, once saltpeter became appreciated for its role in making gunpowder, and thus making war, all nations wanted to secure a supply. For some time after the European introduction of gunpowder, the question of an adequate supply of saltpeter was a universal problem. Trade routes and transportation in the mid centuries were tenuous at best, and the lengths to which nations would go to provide an indigenous supply might seem comical to anyone who does not remember the uranium boom of the 1950s. Today, strategic stockpiles of animal waste may seem bizarre, but what nations did then was merely attempt to mimic nature in making natural saltpeter deposits.

SALTPETER IS WHERE YOU FIND IT

As luck would have it, in that equestrian era there existed an in-place source of saltpeter awaiting exploitation in the form of encrustations on the walls of cellars and stables. In England, special agents of the crown, known for being a rowdy and undesirable lot, were appointed to seek out and fetch these deposits for His/Her Majesty, no doubt the point in time where "s—t detail" entered the language. Once it was determined that these evaporative deposits were the end product of decomposing nitrogen-bearing organic material, "niter beds" or "niter plantations" were built all over Europe, and the petit agents then had to find employment other than as H.M. duly commissioned dung beetles. These niter beds were very much like compost piles, but built on a clay or wood floor. They were open to the air but sheltered from the rain. Layers of decaying animal material (manure, flesh, bones, urine) were built up with layers containing some form of lime (mortar rubble, soil, ashes) and periodically dampened with blood and/or urine. After two or three years the mass was leached for the saltpeter that had been formed by interaction of the various nitrogen compounds and other materials.

In Prussia, farmers were required to build fence rows in such a manner, these being periodically torn down and leached for the saltpeter. In Sweden, until well past the mid-nineteenth century, farmers were required to pay a portion of their taxes in this material, no excuses and no cash

Farrow's Military Encyclopedia. *Volume I., Page 808.*

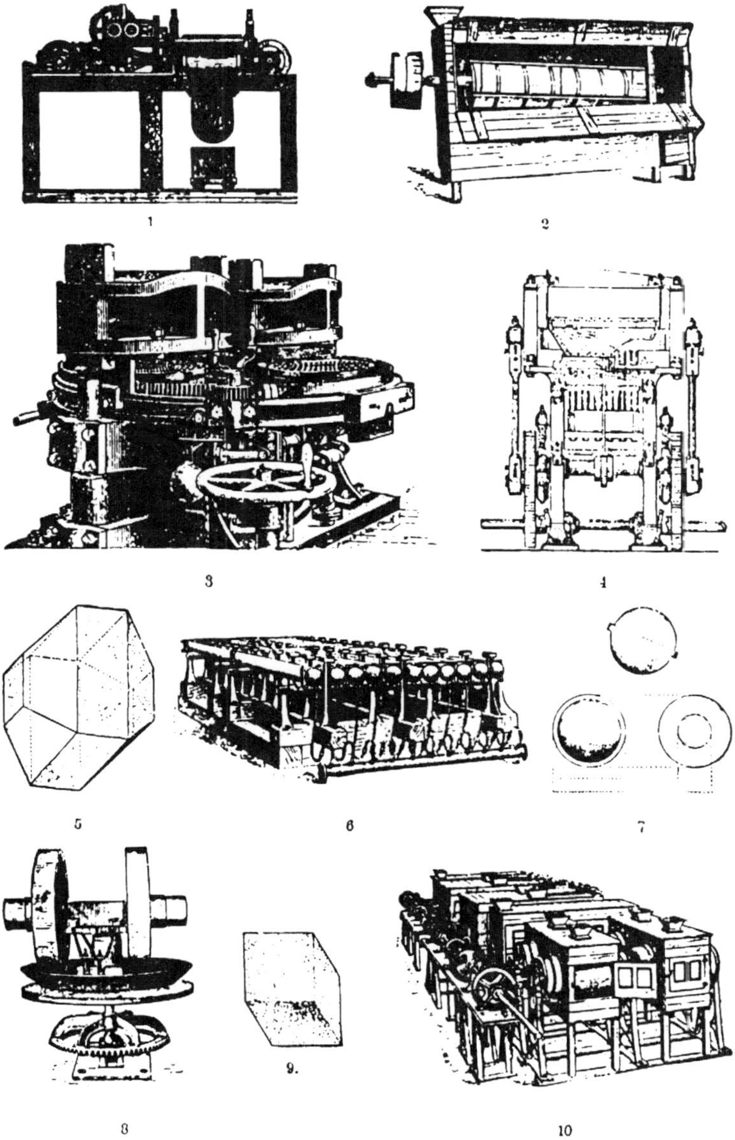

GUNPOWDER.—*1, Pebble Powder Machine. 2, Sifting-reel. 3, Pellet Powder Machine. 4, Prismatic Powder Machine. 5, Schaghticoke Powder. 6, Drying-stove. 7, Compensating Powder. 8, Saltpeter Apparatus. 9, Cubical Powder. 10, Dusting-reels.*

accepted in lieu of same. In England, wood ashes were added to the leaching process to aid in converting any calcium nitrate to potassium nitrate. Records from the eighteenth century indicate that a cubic foot of the good stuff could be leached to produce approximately 4 or 5 ounces of saltpeter.

As nitrates were very serious business, dealing as they did with national security, one of the very few instances where the President of the United States is allowed to use military troops athwart the Posse Comitatus Act is where he is authorized "to employ the land and naval forces of the United States to protect the rights of the discoverer [of a guano island] or of his widow. . . ." Large and concentrated deposits of bird guano such as on small island rookeries were a logistical bonanza in the days when heap-leaching a manure pile was the alternative.

The limestone caves of North America represented another bonanza, because the bat guano's nitrogen content reflected the high-protein insect diet of the bats, and the limestone caves provided lime, access to air, and shelter from leaching rains—everything so laboriously handmade in artificial niter beds. Lightning rods were placed in organic nitrogen-rich soil to attract electrical current that was believed to form nitrate radicals, which would in turn react with potassium salts to make saltpeter.

THE OLD WAY

Suffice it to say, there are many roads to London when it comes to finding saltpeter. And whether one starts with barnyard or stable encrustations, digs up old burial grounds, cleans a cave for bat guano, uses soil from the chicken pen or an old outhouse, or constructs a purpose-built nitrate bed to dispose of his dead cats, ripe road kill, and processed beer, the procedures used over the years for winning saltpeter are largely the same and fairly simple.

The source material is leached with hot water to dissolve

the soluble salts, then the saltpeter is selectively removed from this pregnant solution.

The crudest and simplest method is to put the earth in an iron kettle, cover with water, and boil for a day, stirring constantly during the first couple hours. Floatable organic debris and scum will rise to the surface and are skimmed off. Continue boiling throughout the day, and at evening put out the fire and let all mineral matter settle. The saltpeter solution will tend to be above the settled matter, and it may be drawn off with a siphon (old timers had separate spigots at different levels on the boiling vessel to selectively draw off different portions—not a bad option to suck-starting a siphon hose from such an awful brew).

Draw off this center liquor, being careful to exclude any scum from the top or sludge from the bottom. Reboil this solution until small salt crystals begin to appear, and scoop them out with paper or fine cloth on a wire frame. Discard these. When no more salt crystals appear, set the solution aside in smaller containers. As this supersaturate solution cools, predominately white crystals of potassium nitrate will start to form. When no more crystals appear, collect the crystals, dry them, and rinse them over a screen with cool water. Once dry, they are ready for use. Save this rinse water as feed for the next batch, as you will have dissolved some saltpeter during the washing process.

THE NEW WAY

The chemistry of what is done has not changed over the past several hundred years, but the somewhat more enlightened *way* it is done in a field setting has. Once again we will borrow techniques from the lab coats who labor in abandoned salt mines to devise methods of mayhem for field operatives, as field conditions parallel survival situations.

The supplies and materials you will need are:

- approximately 3 1/2 gallons of nitrate-bearing soil or

other material, screened free of organic debris and pebbles and lumps broken up

- 1/2 cup of fine, sifted, white, hardwood ashes
- a 5-gallon bucket or similar watertight container
- two pieces of finely woven cloth, each slightly larger than the bucket
- a shallow pan or dish, at least as large as the bottom of the bucket, 1 1/2-gallon capacity
- a shallow heat-resistant container (ceramic, metal oil pan, glass pie pan, etc.)
- 1 3/4 gallons of water
- an awl or punch suitable to make drain holes in bucket
- 1 gallon of alcohol (spirit duplicator fluid, rubbing alcohol, whiskey/vodka, etc.)
- a heat source (stove, hot plate, campfire)
- stiff, water-resistant paper (manila folder, etc.)
- porous, water-resistant paper (coffee filter, reinforced paper towels)
- duct tape or rubber band cut from an inner tube

The extraction procedure is as follows:

1. Punch holes in the bottom of the bucket to make a sieve and place one piece of cloth over the holes inside the bucket.

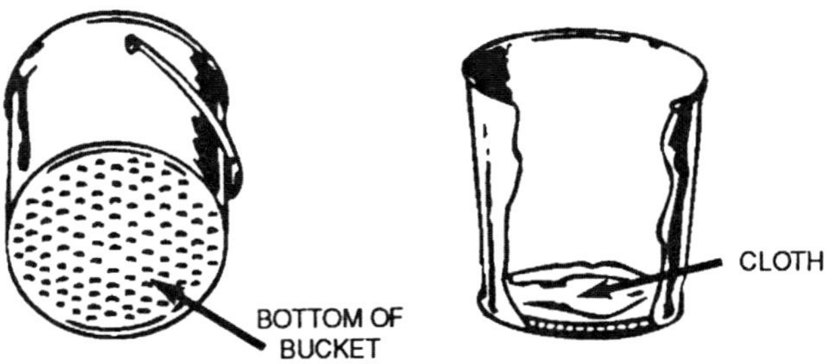

2. Spread the wood ashes evenly on the bottom cloth and cover with the second piece of cloth.

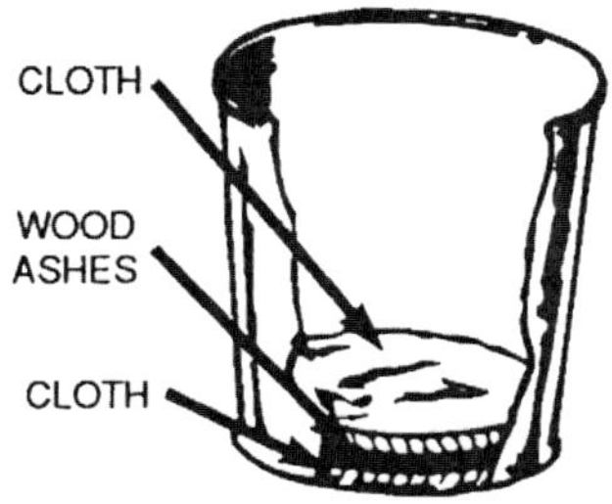

3. Place the nitrate-bearing soil in the bucket, filling to no more than 1 inch from rim.

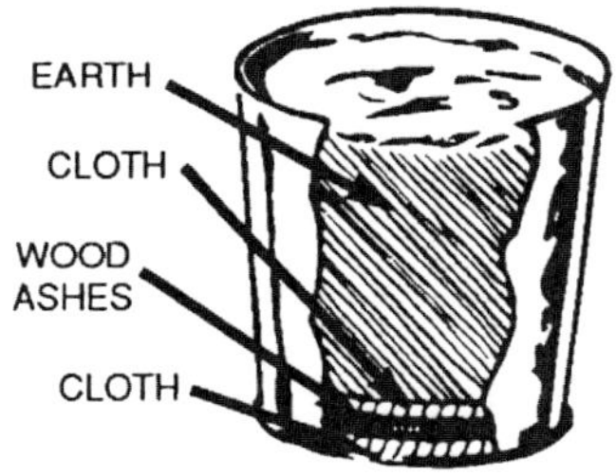

4. Fix the bucket over the shallow container by hanging or supporting it on sticks.

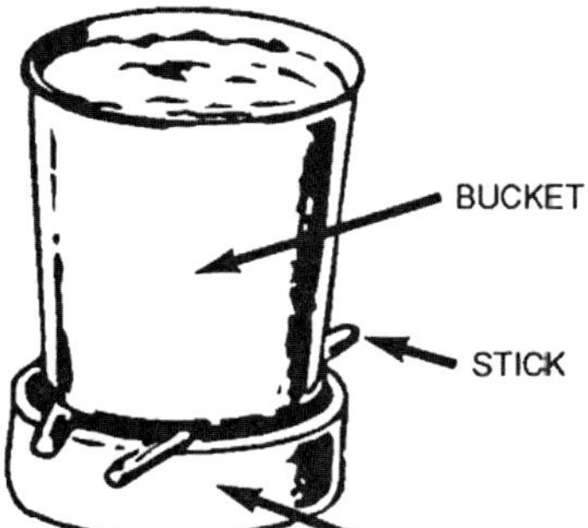

5. Boil the water (a metal plant-watering can with a spraying spout or a tea kettle works best) and pour it over the soil in the bucket—*slowly, a little at a time*—taking care to distribute the water evenly so it percolates through all the soil. Do not pour the water all at once, as this can cause the water to form channels through the soil, leaving some of it unleached.

6. Allow the drained solution to cool and settle for one or two hours, then carefully drain the liquid from the pan into the heat-resistant container, being careful not to pour out any sludge in the bottom of the pan. Discard any sludge.

7. Boil the mixture in the heat-resistant container for at least two hours. Small grains of salt will begin to form in the solution. As they appear, scoop them out with the cardboard or a piece of tight cloth or felt on a wire frame, a coffee filter over a small screen strainer, etc.

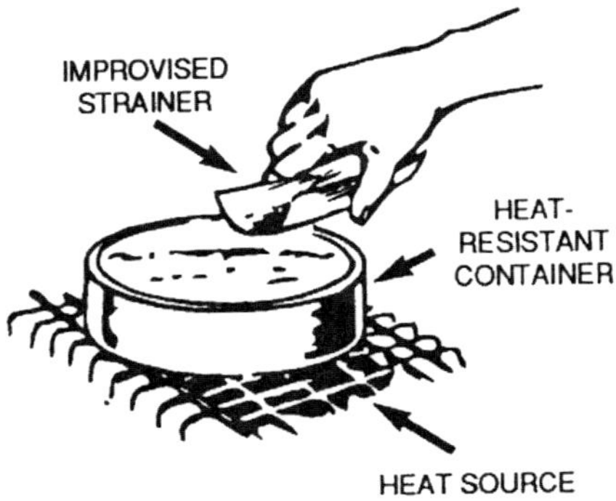

8. When the liquid has boiled down to approximately half its original volume, remove from the fire and let sit for half an hour. Then add an equal volume of alcohol and stir briefly (half a dozen stirs with a stick).

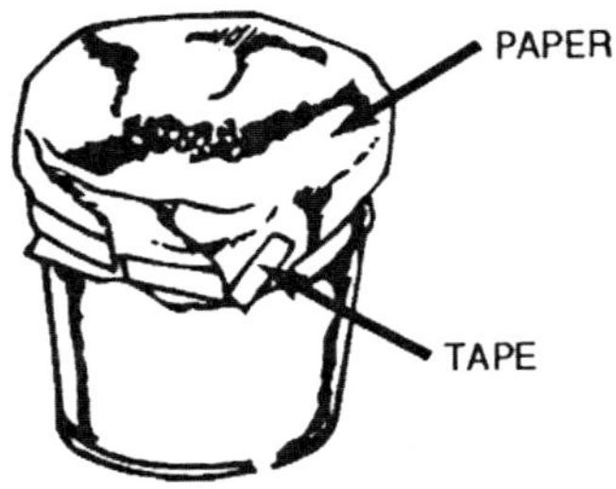

9. Affix the paper filter over the shallow container with tape or rubber band and slowly filter this mixture through it. Small, white crystals of potassium nitrate that have been precipitated by the alcohol will gather. Collect and dry them.

10. To further refine the potassium nitrate, redissolve the dry crystals in the least possible quantity of boiling water. Remove any salt crystals that form as in step 7, and pour the solution through an improvised filter of several pieces of paper, paper towels, or three coffee filters.

11. Collect the concentrated solution and air-dry or gently heat it to dryness. When nearly dry, spread the crystals evenly on a flat surface. When completely dry, the potassium nitrate is ready for storage or use. Do not store in the vicinity of flammables or combustibles such as sugar, flour, sawdust, metal powders, or the like.

The final yield of saltpeter will vary dramatically, depending on the content of the original nitrate-bearing soil. The usual range is 1 to 10 percent.

Chapter 5

CHARCOAL

The easiest component to access is charcoal, which is the primary "fuel" component of black powder. Almost any charcoal will work, and most have been tried, but some are much better than others. The best wood, and therefore most widely used, is dogwood. Because it could not be grown successfully in England, during the nineteenth century the British purchased it from Prussia at $60 to $75 a ton for use in making musket powder. Locally grown willow and alder were thought suitable for cannon and blasting powders.

Whether the species be dogwood, willow, alder, or poplar, wood about ten years old is preferred. In willow or alder this is commonly about 4 inches in diameter, but in dogwood it is about 1 inch. The wood is harvested in the spring when the sap is flowing, only because the bark is then easily removed. When every trace of bark has been removed, the wood is stacked in alternating layers and air dried for two or three years. Experiments have shown that it can be kiln-dried, but it is most important that it have less than 10-percent moisture content when it is charred.

The proportion of charcoal can be increased in powders to be used for blasting, and such powder is *usable* in small arms, but it is slower and leaves a great deal more residue. Powdered coal was attempted as an adulterant during the Civil War by unscrupulous contractors, with deleterious effect. Pit-burned charcoal is preferable for use in fuses and pyrotechnics, but retort-made charcoal is preferred for blasting and propellant powders.

PIT-BURNED CHARCOAL

Pit-burning was the process used to manufacture most of the charcoal of commerce in the old days and, as you would guess from the name, it has nothing to do with a pit. The dried wood is stacked on end in a cone-shaped pile, leaving a hole in the middle. The outer layer is covered with 4 or 5 inches of straw, then ashes, then sod (grass side in), and then covered with an inch or two of loam. Half-inch holes are poked around the bottom, and a quantity of lighted coals dropped down the center hole to start the process. Once the stack inside is burning, the top hole is plugged.

The holes at the bottom are opened or plugged to regulate the amount of oxygen; you only want sufficient oxygen to sustain the fire enough to continue roasting flammable gases and volatiles out of the wood. You want to completely char but not burn up any more of the wood than is necessary to char the rest. Progress of the charring is checked by lifting the sod at the bottom, and if you want the heat to come to any quadrant, you leave those holes open. When the wood is all charred in one area, leave those holes plugged. Such a pile about 10 feet across the bottom and 7 feet high will take about 70 to 80 hours to char. When the

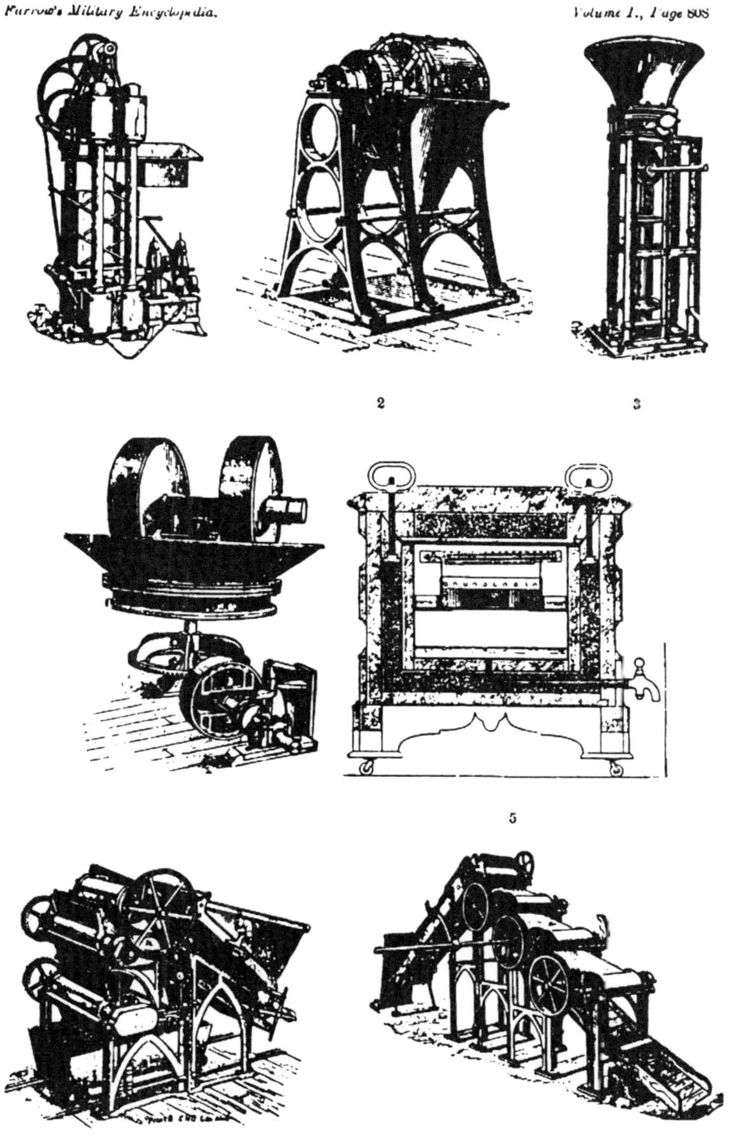

GUNPOWDER.—*1, Powder Press. 2, Mixing Machine. 3, Charcoal Grinder. 4, Incorporating Mill. 5, Hygroscope. 6, Breaking-down Machine. 7, Granulating Machine.*

whole is properly charred, all holes are sealed tightly to stop the combustion process and the pile is allowed to cool.

OR A SNAPPY RETORT

Charcoal for gunpowder is best made in a retort, which is simply some sort of iron or ceramic box that can be heated to the char temperature of wood, approximately 400 degrees Fahrenheit. The box must be vented to permit the escape of gases and steam driven from the wood during the charring process, but closed enough that oxygen cannot enter and burn up the contents. In the heyday of black powder, retorts were large wrought-iron affairs into which wagon loads of wood could be packed. They were arranged over a brick firebox, and the flammable gases and volatiles driven from the wood were vented down to this firebox to aid in heating the wood to continue the charring process.

In a survival situation, a GI ammo can—large or small, depending on how much charcoal you want to make—with the rubber seal and interior paint removed, works fine. It probably won't be airtight with the seal removed, but drill a small hole on the top for safety anyway; considerable flammable gas is generated and you don't want pressure to build up. Immerse this in a hot fire and heat it until no flammable gas can be detected at the vent. It will take several hours. If nothing else is available, you can crimp in one end of a coffee can and insert it tightly in another that has been filled with wood, poke a couple small holes in it, and put it in your campfire.

Good quality charcoal will be clean, crisp, and have a velvety texture and an almost ceramic/metallic appearance. The traditional procedure was to let the charcoal sit for two weeks to let it cool before pulverizing, but it can take as little as a day or so. Be very careful to avoid contaminating it with ashes or to let it absorb moisture from the air while it's cooling. It was usually pulverized in rotary mills similar to an overgrown coffee grinder, or in ball mills with zinc balls. It may be pulverized in the field by shaking it in a coffee can one-third full of small, round rocks, or by grinding it in a dish with a hardwood pestle. The handiest household device is a little electric spice mill, but don't get caught by your wife. It should be ground fine enough that no individual grains can be felt with the fingers, about like powdered sugar or cornstarch. You can sift it through a nylon stocking. Don't let her catch you doing *that* either.

Charcoal must be stored tightly sealed and in small quantities. Not only will it readily absorb water from the air, it will adsorb myriad impurities and pollutants. Many fires in old powder houses were attributed to freshly milled charcoal absorbing oxygen from the air and igniting spontaneously.

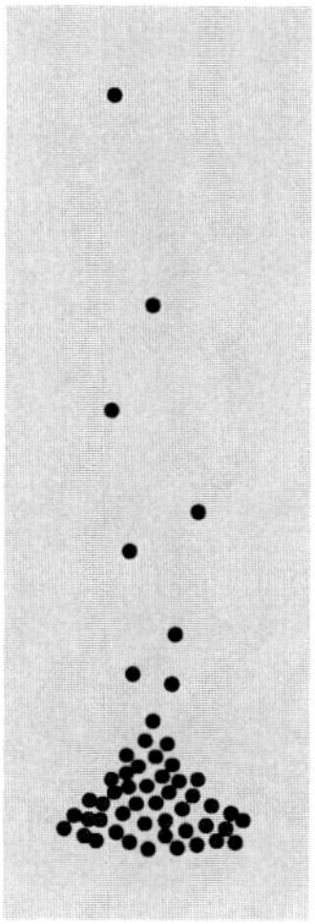

Chapter 6

SULFUR

Sulfur is the least component of black powder, and if you are fortunate enough to be in an area where it occurs naturally, it can be one of the easiest to obtain. It occurs as massive underground and small surface deposits in Texas and Louisiana. Sulfur is mined underground there by injecting superheated steam that melts it, then pumping the molten sulfur to the surface (the *Frasch process*). It also occurs as lumps, nodules, or discrete inclusions in many coal deposits. Low-grade coal deposits in Montana and Oregon often have large-size nodules that can be broken out with a hammer. If you know an old-timer who knows where sulfur pits are, you're in business. If you have access to sulfur-bearing coal, you can remove it as fuel producers do by powdering the coal and floating it out (see technique under "Fool's Sulfur" on page 46).

NATIVE BRIMSTONE

States that have volcanic activity, such as California, Oregon, Washington, Alaska, and Hawaii, produce "flowers of sulfur" (sublimed sulfur deposits) that condense and can be collected at volcanic vents, and at fumaroles such as in Yellowstone National Park. Native sulfur also is found in the rhyolite tuff at Sulphurdale (Beaver County), Utah, at Sulphur Bank (Lake County), California, and on Lassen Peak, California.

If the sulfur is mixed with gangue in proportions too lean to simply be melted out, the host rock can be crushed/ground and the native sulfur then sublimed/distilled out in a ceramic or iron retort. Elemental sulfur can also be floated out of finely milled host rocks, as are sulfide minerals (see below). In many states, groundwater contains hydrogen sulfide (H_2S) gas, which leaves the water when it is allowed to sit, and may be captured in a gas bag and used as outlined below. Some such water contains enough H2S that it will actually convert to elemental sulfur and settle out after sitting for a while. Many of these wells, however, especially in the Northwest, have a lot of other gases, minerals, and dissolved salts. Many of these minerals and dissolved salts stay in solution only under pressure, and these will also precipitate when the water sits at surface pressure. When this is the case, you have to then distill/sublime the elemental sulfur from the mass of precipitates, or only capture and use the H2S as outlined below.

One could, of course, lay in a stockpile of sulfur, but if you're going to the trouble to do that, you might just as well lay in a supply of musket powder. When buying sulfur, the cheapest will be found at the garden or agricultural supply store. Chemical suppliers and drugstores will also have it, in greater purity at greater cost.

It's a longer way around the bush, but almost every state has deposits of some sort of sulfide ore that can be used as a source of sulfur.

Farrow's Military Encyclopedia. *Volume I., Page* 808.

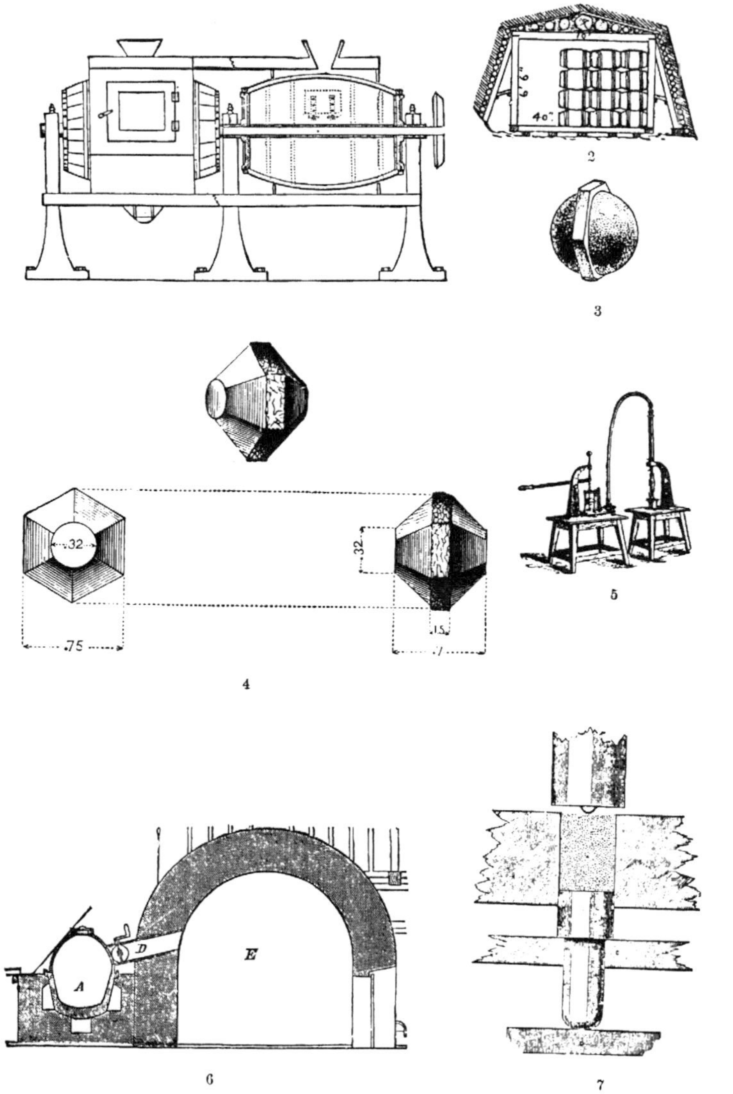

GUNPOWDER.—*1, Glazing Apparatus. 2, Storage of Gunpowder in Magazine. 3, Sphero-hexagonal Powder. 4, Hexagonal Powder. 5, Densimeter. 6, Sulphur Refining Apparatus. 7, Special Powder Machine.*

FOOL'S SULFUR

Winning sulfur from sulfide ores starts with gathering the ores. In many mining districts there are tailing piles that contain almost pure sulfide ore, abandoned after the valuable minerals were extracted. Natural rocks containing sulfide inclusions also can be milled down with a hammer and hand-panned for the sulfides. The most common sulfides are iron pyrites ("fool's gold") that vary in color from golden yellow to silvery, sphalerite (zinc sulfide, grayish color), galena (lead sulfide, shiny silver color), chalcopyrite (copper sulfide, golden to polychromatic), and arsenopyrite (arsenic/iron sulfide, gray to olive color). There are hundreds of distinct sulfur minerals. All are heavy enough to be hand-panned once milled free of the host rock, and all can be used as a source of sulfur.

When looking for sulfide ores, watch for weathered rock that has red mineral stains; if the surface shows tiny red-lined pits where sulfide crystals have weathered out, dig on in. Once past the oxidation zone, if you find sparkly crystals from yellow to olive, and if they crush under a knife point rather than flake apart like mica, odds are good you have a sulfide. Sulfides will also collect in the stream beds of sulfide-mining districts, as do black sands (their density is almost exactly the same as magnetite black sands), where they often can be gathered with a shovel. Iron pyrite is also a common contaminant of Midwestern coal deposits. Roast the suspected sulfide over a hot flame and you will be able to smell the unmistakably pungent aroma of sulfur.

Pyrites have often been exploited in times past by burning to produce sulfur dioxide (the basis for making sulfuric acid) and using what remains as iron ore. During the Depression era, many schemes were developed for removing sulfur from pyrites, coke-oven gas, smelter fumes, and hydrogen sulfide gas. This by-product sulfur was produced in Canada, Germany, Norway, Portugal, Spain, Sweden, and the United States. Because of development of the massive underground deposits of elemental sulfur along the Gulf Coast, these sulfide resources have not been widely used except as a source of sulfur dioxide, but serve to show how common sulfur is, especially in sulfide form.

Garden stores often have bags of sulfide mine tailings for sale as fertilizer, used to lower the pH of soil, add iron, and so forth. Most I have examined contained a high percentage of arsenopyrite, so nobody in his right mind would put them on his tomatoes, but they do represent a possible source of off-the-shelf sulfides.

If one has even a rudimentary lab setup, there are many ways to win sulfur from sulfur compounds. But the *Claus process* that is used to win sulfur from sulfides and from smokestacks is adaptable to a bare-bones survival situation, where operational and equipment options are severely limited. When pyrites are burned or roasted to produce sulfur dioxide gas (SO_2), this is captured in water, where it forms sulfurous (H_2SO_3) acid. When hydrogen sulfide gas is introduced into sulfurous acid, it forms sulfur almost instantly. This is the heart of the Claus process, now used primarily to clean hydrogen sulfide and/or sulfur dioxide gas from various industrial emissions. In essence, what we will do here is create SO_2 and H_2S emissions on purpose from scroungeable sulfur-bearing materials, then adapt the Claus process to reclaim the sulfur.

SULFUR FROM SANTA CLAUS

Both sulfur dioxide and hydrogen sulfide, of course, are soluble in water. If you have a little citric acid to add to the water, it will enhance things. If a solution of hydrogen sulfide alone is left exposed to the air, sulfur gradually forms and can be filtered out. Hydrogen sulfide can be produced by the destructive distillation of coal that contains sulfur (most does), or by putting hydrochloric acid (HCl) on metallic sulfides such as ferrous sulfide (FeS), which gives the reaction $FeS+2HCl \rightarrow FeCl_2+H_2S$. In the event you do not have HCl, see below for other ways to generate H_2S.

Bear in mind that hydrogen sulfide (yes, the one you made in high school for its rotten egg smell by heating sulfur and wax) is poisonous (ask any coal miner) and flammable. Always work with positive ventilation (a fan exhausting to the outside). Sulfurous acid (H_2SO_3), although not as active as sulfur*ic* acid (H_2SO_4), can cause damage to clothes, eyes, and skin.

If our raw material is sulfide ore of some sort, the first thing to be done is concentrate the sulfides. Ore must be broken down to about the size of table sugar to do this. This may be the point when you consider leaving your musket on the rack and building a bow and arrow, as anybody who mills ore with no more than a hammer earns his pay. If all you have is a hammer or two rocks, then work on a plastic sheet or tarp to catch the flying pieces, and wear eye protection.

If you have a survival shop with tools, a chain mill can be built from a brake drum and salvaged components (see Chapter 8) that will come in handy for all the various milling operations. A concrete mixer with fist-sized hard river rocks (about 1/3 full of rocks, plus enough nut-size broken ore to fill the spaces between, and enough water to make a very soupy mix) will also serve as a noisy but effective improvised ball mill.

Once the ore is broken down to sugar size or finer, the sulfides must be removed from the gangue (the stuff in the host rock you don't want). You can do this with a hand pan, like you would pan out black sand concentrates and gold. Given a survival shop, you can improvise a flotation cell, which is used in industry to separate milled sulfides from ore and is amazingly efficient.

A flotation cell simply is a vessel with a violent agitator

(I've seen them made from old-style washing machines) and a way of introducing air into the bottom of it. The sulfide ore, water, and kerosene (or diesel, or pine oil, or any of a hundred specific commercial flotation agents) are agitated with air. The oil and air form a froth with the sulfides, which floats to the top and is skimmed off. Rinse this foam and you will have fairly pure sulfides. Industry uses sophisticated, continuous-flow cells with reagents designed to float specific sulfides and reject others, but for our purposes a batch-type cell will work fine because we want any sulfide we can get, since our end product is to be sulfur.

An absolute minimalist float cell would be a coffee can with a plastic lid containing about 2 inches of milled ore, water 2 inches over that, and a quarter teaspoon of kerosene. This is shaken violently and long, then left to settle so you can skim off the sulfide-carrying foam. A concrete mixer can also serve as a float cell; let it run at top speed for several minutes with a very soupy mix, shut it down, and skim off the foam, repeating until you have all the sulfides. Other options for the flotation of sulfides are given in Chapter 8.

You can also gravity-concentrate milled sulfides from gangue material with a little gold sluice or rocker box, a shaking table, or any of the wonder devices sold to weekend prospectors such as spirals and wheels.

Once you have the reasonably concentrated sulfides (they don't have to be pure for our plastic-bucket-and-shovel chemical engineering), the sulfides are roasted in a retort to drive off the SO_2 gas, which leaves via an exhaust tube and is captured in water. The temperature at which all the SO_2 is driven off will vary with the ore, but when you first open the retort, if the sulfides still look like sulfides you will know to roast them either hotter or longer, or both, next time. A good starting point is to roast them for a full hour at cherry-red heat. This means if you're doing it over a wood fire, lay in plenty of good, dry fuel, and build the fire in an enclosed "fireplace" where the retort is surrounded by coals. If you can improvise any sort of blower to increase the heat of the fire, you're that much ahead.

Mild steel makes a suitable retort, and the easiest one is made from pipe fittings. Follow the schematics given here, and adapt your design to fit what materials and components you have. Assemble the pipe fittings loosely (so you can get it back apart), but since the retort should be airtight so you do not lose the SO_2, lute the outside of the fittings with clay. A little graphite (ground pencil lead) on the pipe threads when you assemble them will help them come back apart.

There are two important safety considerations when retorting sulfides. First, bear in mind that many unhealthy heavy metals (lead, cadmium, arsenic, etc.) may be in the sulfide ore and can also volatilize, so work with proper ventilation. Secondly, when exhausting any retort discharge tube into water, remove the exhaust tube from the water before you stop heating the retort; if the retort cools sufficiently to draw water back up into itself through the exhaust tube, it can instantly make steam and explode (not an uncommon occurrence with the old mercury retorts gold miners used).

After all your hard work, at this point you have a vessel of SO_2 dissolved in water (sulfurous acid, H_2SO_3). Generate hydrogen sulfide gas by the method of your choice and introduce it into the vessel of sulfurous acid, and you're in the sulfur business. But that's just one option.

999 BOTTLES OF SULFUR IN THE WALL

Probably the most massive occurrence of mineral sulfur in America is as gypsum (calcium sulfate). It is open-pit mined in numerous locations, and in Florida more than a *billion* tons of phosphogypsum—produced when sulfuric acid is used to make phosphoric acid from phosphate rock during fertilizer production—will have been stockpiled by the turn of the century.

You don't even have to leave your own house, however, to find gypsum. It is the base for your Sheetrock or drywall plasterboard walls and ceilings, and for the plaster-of-paris flamingos and baby deer that suburbanites are wont to mount in weird juxtaposition on their front lawns. And, if you care to sacrifice it, the alabaster statue Aunt Maude brought back from Italy is just a harder form of calcium sulfate.

Until recently, not much thought was given to removing sulfur from gypsum, but the underground deposits amenable to extraction via the economical Frasch process are waning. The looming prospect of a billion tons of stockpiled by-product phosphogypsum spurred the U.S. Bureau of Mines to take a look at how to use this gypsum for its sulfur content. Although sulfur is the thirteenth most common element, world reserves of easy sulfur will have been depleted significantly by the year 2000, and the sulfur in this phosphogypsum represents considerable value. To a survivalist, gypsum is worth consideration as a source of sulfur because it is all around us—odds are good it's over your head as you read this.

To keep processes within minimalist bounds, we have adapted current Bureau of Mines research to convert the gypsum to calcium sulfide, then used the *Elcor process* to release the sulfur from the calcium sulfide as hydrogen sulfide gas. The hydrogen sulfide can then be converted to elemental sulfur by dissolving it in water and letting it sit, or introducing it to a sulfurous acid solution, where it instantly converts to sulfur. If you only have gypsum to work with, the H_2S produced can be burned in a closed chamber with gently forced air to make SO_2, which is exhausted into and captured by water to make the sulfurous acid you also need.

Given a processing plant, there are shorter ways around this chemical bush, but in a survival lab (the barn or woodshed at best) we have to stick with ambient-pressure wet chemical methods utilizing reagents we can expect to pull out of the ground with hand tools or off your living room wall.

To convert gypsum to calcium sulfide, it is finely ground and mixed with powdered soft bituminous coal (40 percent) and powdered magnetite (10 percent). To get the magnetite, you can pan out river sand until you get the heavy black sands or use any of the other gravity techniques mentioned above for concentrating sulfides, then use a magnet to pull out the magnetics, which will be nearly pure magnetite sand. In this reaction, the carbon of the powdered coal steals the oxygen from the calcium sulfate (gypsum), and the magnetite is merely a catalyst to help the reaction take place at a lower temperature. If no coal is available, the substitute is powdered charcoal (see above), but the reaction takes longer. Do *not* do any other substituting or improvising with this formulation. If, for instance, you decided to substitute aluminum for the coal to use up oxygen, and substituted red iron oxide (ferric oxide, or rust) for the magnetite (ferrous or black iron oxide), you'd have built yourself an incendiary bomb that generates a temperature of more than 4,000 degrees Fahrenheit.

These components are mixed thoroughly and then heated in a retort (as you used to roast the sulfides). They should be heated to 900 degrees Celsius and held at that temperature for thirty minutes. At 900 degrees Celsius (equivalent to 1,652 degrees Fahrenheit), steel is cherry red. As some sulfur dioxide is produced in this reaction, it is worthwhile to vent the retort into water to catch it. Remember to remove the exhaust tube from the water before removing the heat source so it won't cool, suck water into the retort, and generate a steam explosion, as we warned about earlier. Even if you opt not to capture the sulfur dioxide generated in this step of the process, *you still must vent the retort.* Any retort loaded with any materials that will generate a gas *must* have a vent hole. An unvented retort is not merely a retort—it is a pipe bomb.

Once the calcium sulfate has been converted to calcium sulfide in the above reaction, the contents of the retort are reground if necessary. It is important this calcium sulfide be finely divided because it is sparingly soluble in water, and for the next step it is necessary for as much of it to dissolve as possible. The calcium sulfide is then actively mixed with water and carbon dioxide (CO_2) as a slurry in a reactor vessel.

Since we are working at atmospheric pressure and merely guiding gases in and out of this reactor vessel, building one is fairly simple. A stainless pressure cooker (or cast iron—no aluminum unless you coat it with epoxy) works fine. The lid will have a hole for a pressure release and usually for a gauge. In the hole in the center of the lid (it could be either), insert a rubber bushing (such as used where electrical wires go through a metal box). Grease the hole in the bushing and insert your mixing rod through it. You can also use a rubber lab cork of the right size, also greasing the hole. The stirring device can be something such as a paint stirrer in an electric drill. If you have a properly fitting lid on your improvised float cell (above), it can do double duty as a reactor vessel if you fit a stirrer and feed CO_2 in the bottom instead of air

In the other hole in the lid (if there isn't one, then make one), seal a plastic hose to carry the H_2S off and into the sulfurous acid. Then at the bottom center of the pot, seal in a small (1/4-inch or less) plastic tube, which will be used to carry CO_2 *into* this reactor vessel. While this calcium sulfide/water slurry is being agitated, CO_2 is introduced into the mix through this tube in the bottom. The calcium would rather be a carbonate than a sulfide, so it reacts with the CO_2 to make calcium carbonate (limestone) and throws off H_2S gas. This H_2S is vented off the top of the vessel and goes over into the sulfurous acid solution, where it forms elemental sulfur.

The CO_2 you need can be generated in many different ways. You can mix baking soda and vinegar, mix hydrochlo-

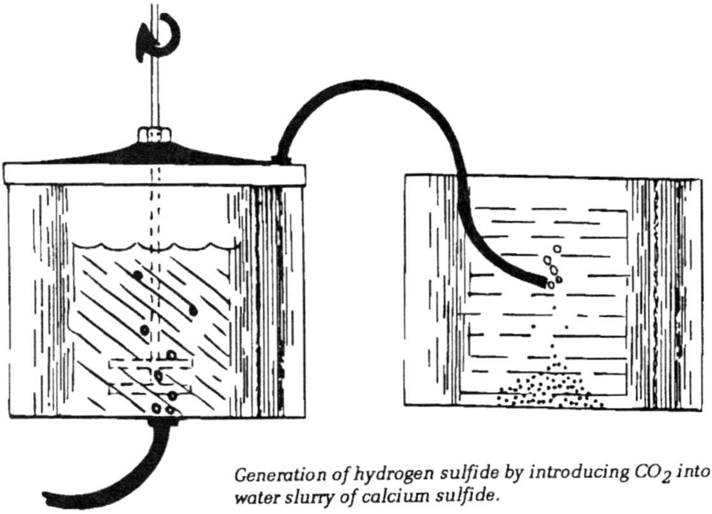

Generation of hydrogen sulfide by introducing CO_2 into a water slurry of calcium sulfide.

ric acid and ground limestone, use a seltzer bottle, put dry ice in a closed container and melt it, make wild fruit vinegar and react it on powdered limestone, give very small blasts from a CO_2 fire extinguisher . . . there are innumerable ways; use what you have. The vessel you generate CO_2 in should be airtight so the gas will go out of its tube and into the reactor vessel under a pound or two of pressure (remember, it's going in at the bottom of the reactor vessel), but not something that will contain excess pressure until it explodes. To make sure you do not build up unsafe pressure, it may be a good idea to drill a 1-inch hole in the lid of the CO_2 generator and close it with a cork that will pop free under a few pounds of pressure.

Do not generate CO_2 too fast or it will just blow on through the reactor vessel and be wasted. One way to regulate the CO_2 feed without resorting to a pressure vessel with valves and so forth is to use dry components (such as baking soda and citric acid, baking soda and aspirin, baking soda and vitamin C) and meter the water into the CO_2 generator through a valve; when you need a little more CO_2, meter in a little more water until the reagents are all used up. You could also mix suitable reagents (or blow CO_2 from a fire extinguisher) into a pressure vessel with an exit valve, such as a compressor tank. And, of course, alcohol or natural gas will burn to produce CO_2, so you can even burn up all the air in a gas bag of some sort and pump that through the reactor vessel; the 70 percent or so of this depleted air that is mostly nitrogen will not interfere with the reaction of the CO_2. Any time you improvise, carefully think through what you are doing. Be careful not to build a pressure vessel inadvertently unless you know what you are doing in this area.

After all this hard work, you now have sulfur from Sheetrock. The sulfur formed will be very fine and contain a certain amount of acid. To use it for gunpowder, it is filtered out, then washed with clean water in the filter. When washed and dried, it is remelted in a covered glass or ceramic dish (remember, sulfur will start to sublime away before it melts), then finely reground to give the better form of sulfur to use in gunpowder.

A DOSE OF SALTS

If you don't want to bust a hole in your Sheetrock wall, the process given above can be used on the sulfates of some alkali or alkaline earth metals by converting them to their sulfides by the above process, then robbing them of their sulfur.

In some settings this can be a viable source of sulfur, as most of the "alkali" flats or lakes in the Southwest are composed of mixed sulfates and chlorides, mostly of the alkali metals (because they react so readily in nature). They are, therefore, potential sources of sulfur for a scrounging survival chemist. When these mixed salts are treated as above, most alkaline sulfates convert to less-soluble sulfides, which you want, and the chlorides that are very soluble are then separated from the sulfides by washing with cold water.

Look for alkali flats/lakes with a particularly soapy feel. Calcium sulfate, being sparingly soluble in water, can even be reclaimed from seawater, but the process is too tedious to be practical even for survival sulfur production (unless you are near a place where seawater is evaporated for salt, since such operations often end up with a pile of gypsum).

BUG SPRAY AND BATTERIES

If you have any lime-sulfur spray, as used to control pests and fungus on fruit trees, filter it to remove any free lime, then introduce hydrochloric acid to precipitate sulfur from the remaining calcium polysulfide solution. Then wash this sulfur in a filter funnel to remove any calcium chloride. Just be *sure* you are starting with lime-sulfur spray only. Introducing hydrochloric acid (or any other chemical reagent) to an *unknown* pesticide is deadly business.

If you have no caveman sulfur resources and no Sheetrock, the battery (sulfuric) acid in abandoned cars can be used as a sulfur resource. Hot, diluted battery acid (from a fully *charged* battery) can be used to digest limestone to make calcium sulfate and give you a gypsum source that is manipulated as outlined above to coax the sulfur out as H_2S. Always wear gloves and eye protection when working with sulfuric acid, and have some baking soda handy (already in solution in a spray bottle). If you don't have baking soda, use wood ashes dissolved in water to neutralize any spills or splashes. Do not use wood-ash solution in your eyes, as you'd merely be treating a corrosive acid with a corrosive alkali.

Chapter 7

SWEET SOLUTIONS

In the event that you have neither sulfur nor the ability to produce it, suitable propellant powders can be made from saltpeter, sugar, and ferric oxide, as we outlined in Chapter 3. The three items needed to produce red-or-white propellant powder will be available in most settings.

Sugar is a nearly universal commodity, and it is also one that is very simple to produce from nature if there is no other source of supply. It can be produced from a number of plant sources, but the easiest two are sugar beets and maple trees. Sugar cane also works well if you have it, and if you have the patience, corn stalks will yield a sugar-type product, though relatively little in comparison to beets and cane.

Sucrose ($C_{12}H_{22}O_{11}$) is the sugar we specifically want for making red-or-white gunpowder. Cane is crushed, and beets very thinly sliced or shredded. They are placed in warm water, and as the water is heated, a small quantity of lime water is added. The purpose of the lime water is to raise the pH to slightly alkaline (pH 8) to prevent the sucrose from breaking down into simpler sugars. If you do not have litmus paper to measure the pH, add the lime water very slowly, a little at a time. When a white precipitate starts to form, this indicates an alkaline solution; you have enough. The beet slices/cane are then boiled to dissolve the sugar, then the solution strained and filtered. If you have a cider or wine press to squeeze the pulp, they work fine—if not, just twist it up in a cloth or cloth sack and squeeze it as you would to get fruit juice to make jelly.

Maple sap is easily collected from the trees in late winter when the sap is rising by drilling half-inch holes through the cambium layer that carries the sap. The sap is gathered by sticking a tube or miniature trough in the hole up to the cambium layer (an aluminum gutter-spike ferrule works very well) and catching it in a bucket. Depending on the species of maple and the climatic conditions, you can count on having to empty the bucket at least once a day. Many commercial producers simply plumb the maple tree with plastic tubes leading to a collection container.

For our purpose, the procedure from here on is the same for all sugar sources under consideration. The maple sap or filtered beet/cane/corn sugar solution is then percolated slowly through bone charcoal; if you do not have bone charcoal, then use wood charcoal that has been *lightly* sprinkled with wood ashes.

Once filtered through charcoal, the liquor is then boiled down to a supersaturate solution. The sugar of commerce is produced with a more complicated process that delivers an enhanced (more pure and aesthetically snow white) product, but we do not plan to eat what we make here. Commercial processors all boil the solution down by drawing a vacuum; we will simply boil it with heat.

Maple sap usually produces sugar at a ratio of about 42:1; that is, 42 gallons of sap will boil down to produce a gallon of sugar. Sugar beets can contain as much as 20 percent sugar. Sugar cane is also very rich in sugar. Corn stalks and cobs are very lean and yield a dextrose/glucose/maltose product that makes easy syrup but difficult sugar. By the time it is cooked to hardness, I'm not sure what the final product of corn sap is, but it is decidedly a second-choice component to sucrose for making red-or-white powder.

There are two things to watch for when boiling sap or juice down to sugar, and a couple tricks to get you out of trouble. First, once the syrup begins to thicken, it is wont to suddenly rise as a froth in the pot and boil over. To stop this, add just a dash of butter to the pot the second it starts to rise. Since we are not making an edible product, a little bacon grease will do the same thing.

Second problem is, once the syrup has boiled off all the water of solution, it is no longer syrup but melted sugar. If further heated, the melted sugar will begin to caramelize or drive off the water of crystallization and convert to pure carbon. So this is one boiling pot you should watch. Watch for any *sudden* darkening of the syrup (that is, darker than its natural light brown) that would indicate the beginning of caramelization, and immediately remove it from the heat if you see it. Once the syrup has thickened, test it often by dipping a small quantity out of the pot and dropping it in cold water; when it forms hard strings in the cold water, it has "sugared off" and should be removed from the heat. Once cooled, it should harden into a sugarcake. If it does not harden on cooling, then it was not boiled long enough. Simply boil it some more.

This sugarcake is then pounded into small lumps so it can be measured for the making of red-or-white propellant.

MAKING IRON OXIDE

Most tool and gun owners spend a lifetime trying to prevent the formation of rust. In this case, we will make it on purpose. We want red iron oxide (ferric oxide Fe_2O_3), not to be confused with the black iron oxide (Fe_3O_4) that is made by burning steel wool.

If you have fine steel wool, you can make ferric oxide very quickly. Clean the steel wool with solvent (it comes from the factory lightly oiled) and dry it. Fluff it apart and hang it in a closed jar with a small bottle cap full of hydrochloric acid at the bottom, and it will turn to rust almost overnight. When completely rusted, crush it in your hand as you would a shredded-wheat biscuit and sift it through a fine screen.

Virtually any iron or steel object, if perfectly clean, will also rust quickly if exposed to damp air, especially damp salt air, or air with HCl vapor. Simply scrape off the rust with a sharp blade or scour it off with steel wool. Do not use sandpaper or emery, as this will introduce unwanted grit contamination to your ferric oxide.

A gun-eprouvette, used to determine the strength of gunpowder by the amount of recoil produced.

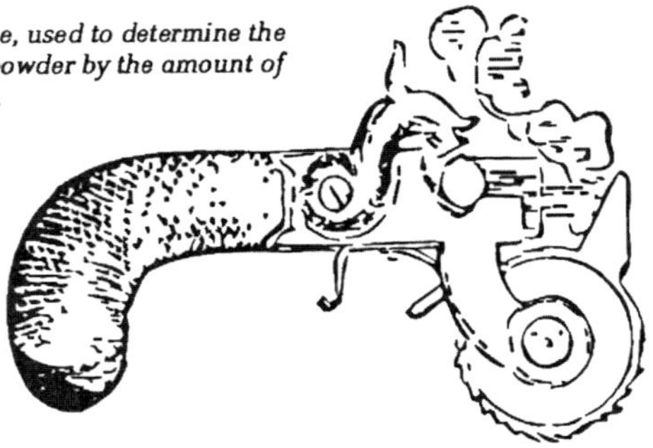

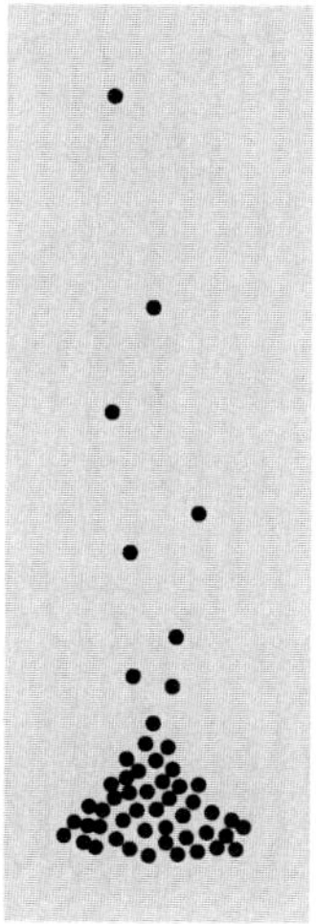

Chapter 8

THE MINIMALIST MINER

Although you may get mad enough at your spouse to tear the Sheetrock off the walls (after you've been crowned for grinding charcoal in her spice mill), winning some of the other resources from nature, dressing the ore, and removing what you want for the project at hand can get a little more complicated. You may surprise yourself at how much hard rock you can mine with no more than a pick and shovel and rock bar, *but that's just the beginning.*

The mining industry is well supplied with innumerable tools and machines that make child's play of dressing minerals in the quantities we're talking about here. But unless you happen to be so unfortunate as to be a miner, in a survival situation you need to know how to grow your own machinery.

Some fairly minimal machinery can be constructed easily from scroungeable components that will give a labor-saving return all out of proportion to the amount of work it took to build them. We'll discuss them in the order you'd probably use them and outline what they are used for. Some will serve more than one purpose.

IRON-BOX HAND CRUSHERS

The first step in dressing ore is to crush it to a size where it is suitable feed for whatever your milling setup is. By the time you hand-mine, you probably won't have a lot of big pieces, but the ore should be half-inch or smaller to be suitable feed for the type of mills we're considering.

SAFETY NOTE: *Puh-leeze read and heed.* Most rock is hard, but not hard to break. Since it has almost no elasticity, when you hit it, it breaks, even when you hit it with a tool that is a lot softer. And when it breaks under impact, it tends to throw little, sharp pieces. And these glass-sharp pieces will, I promise, sooner or later find your eye. So wear safety glasses; squinting doesn't cut it. Murphy will see to it that the fragment with your name on it goes into your master eye. What's the use of musket powder if you can't sight your musket?

"Iron-box" hand crushers are just that: an iron box that contains the rock while you pound it into little pieces with a long-handled pestle. In fact, it doesn't even have to be iron. A wood box from 8 inches to a foot square will also contain the flying pieces. Either box is simply placed on a hard, strong, flat surface (such as a concrete slab, flat rock, or heavy iron sheet). Larger chunks of ore, from the size of a grapefruit down to the size of an egg, are placed in this box a few at a time and pounded into smaller pieces. Pieces about a half-inch through are suitable feed for improvised mills such as we will cover below.

The most effective pounding instrument is a 4-foot length of water pipe with a flat-bottomed chunk of iron welded to it. This piece of iron should weigh between 2 and

10 pounds. Many things work well: a piece of steel shafting, the head of a sledge hammer, a wood-splitting wedge mounted upside down—use what you have. The heavier, the better, because to operate it you simply raise it a couple feet and let it drop on the rock you want to break. Just as with breaking stone *in situ* with a miner's pick, a couple of moderate but well-aimed blows are more effective (and *much* less tiring) than one give-it-all-you-got shot. Use your energy to lift the weight, then let the energy in the falling weight break the rock.

The advantage of using a drop weight on the end of a handle, as opposed to a hammer, is that it keeps you remote from flying rock fragments and keeps you in a much less tiring posture while you work. In effect, you and your "iron box" are working as a one-lung stamp mill.

In the old days, stamp mills were used to reduce ore down to the final size. If you have a smooth surface to work on, you can do it all with an iron-box crusher by scooping out the crushed contents of the box, screening out the fines, then returning the oversize to the box for more pounding. But if you have the ability to construct one of the impact-type mills outlined below, it is more efficient to use your iron box only to break ore down to 1/2-inch feed for the impact mills.

IMPACT MILLING MACHINES

The advantage of these mills is that they are put together readily from scroungeable components, and although they are not industrial production machines, they have no problem milling the ore fine enough to release the sulfides you are after, or reducing magnetite sand or coal or charcoal to talcum-powder fineness.

A rotary impact mill simply is a round chamber that houses a set of rotating hammers, which strike and pulverize any ore in the chamber. Because of the wear factor, such mills are seldom cost-effective for production mining except for pilot mills and weekend you-and-me-Joe operations. But since they will do the job and can be put together cheaply, they do just fine for small runs or when you want to pulverize rocks without spending any money. Essentially, there are two types of mini impact mills.

The Chain Mill

The chain mill is a batch-type mill with no discharge. The mill is buttoned up, then turned on. A small quantity of broken ore is slowly fed into the mill, where a rotating chain flail pounds the ore into a powder. The longer you leave it in, the finer it gets. It is handy for assay work, as a measured amount of ore can be milled in a captive environment. Such a mill will pulverize coal and charcoal in a flash to a powder so fine it is almost fluid—and *very* explosive when exposed to the air. Don't have any spark or flame sources around when you operate or open it up after milling coal or charcoal. And no matter what you are milling—rocks, Sheetrock, black sands, or charcoal—none of it is good in your lungs or eyes. Always wear eye protection and a dust mask.

Scrounging Parts

The easiest junk from which to build a small impact mill is a steel brake drum. They are made from good quality cast steel, and they already have about the proper shape and dimensions. You can get them for little or nothing at your local auto salvage yard.

To build a chain mill from a brake drum, mount the brake drum vertically on a steel framework. Behind it, mount your engine or motor. If you plan to use it a lot, it is a good idea to mount the chain on a jackshaft and drive that with a belt from the shaft on your engine or motor. You can also mount the chain directly on the engine or motor shaft if it is long enough, but the vibration is hard on motor bearings.

The shaft extends through the axle hole in the back of the brake drum to the center of the drum. Depending on the brake drum you use and the diameter of the shaft, you may be able to seal between the shaft and the axle/spindle hole in the drum by merely greasing the shaft and wrapping it with a rag. Or you may have to make an adapter plate that covers the hole in the brake drum and has a hole the size of the shaft. The shaft should be at least 3/4-inch in diameter for very small mills; for anything over a foot in diameter you should go to 1 inch or larger and use good solid bearings—a chain flailing and flopping on the end of a shaft puts quite a strain on it.

The front of the mill is closed with an iron plate that is at least 1/4-inch thick and has a central feed hole and a feeding chute made from 1-inch iron pipe. This feed spout is important for two reasons: without it, the odd rock may bounce back out of the feed hole straight at you, and the milled material will likewise tend to spill out, especially if you try to get too much ore in and "choke" the mill.

This plate can be attached by welding 3/8-inch bolts, which go through holes at the edge of the plate, to the rim of the brake drum. The plate is put over these bolts and held with nuts and lock washers. If you put a bolt every 3 inches or so around the periphery of the brake drum, the plate should seal up dust-tight.

The chain should be of a length to almost touch the inside rim of the brake drum. How you attach the chain depends on the configuration of the shaft you are working with. Welding the chain to a locking collar works as good as anything if you use a longer setscrew on the collar and drill a shallow mating hole in the shaft. This lets you easily take the chain off the shaft for maintenance. You should use as heavy a chain as you have. The end link will wear quickly, and the easiest remedy is to use a shackle, which can be easily replaced, as the last link. All chain or shackles should be free of zinc, as tramp zinc in your pulverized material can get in the way of your chemistry.

Chain Mill Operation

SAFETY NOTE: Never stand at the side of any homemade impact mill when it is running. If the chain breaks or an impact arm comes off, it can go right through the side of the mill and get you. Always stay either to the front or rear of an operating impact mill. Rig your engine/motor with a remote kill switch so you can shut it down from a distance if something comes apart.

To operate the chain mill, turn it on and every few seconds slowly feed ore of a size to readily drop down the feed chute, about a jelly-bean-sized piece. There are two feed considerations: material must be bone-dry or it will gum up, and if you feed in too much it will "choke" the mill. With more than it can chew and keep flying around inside the mill, the chain tips cannot extend, and this puts a great strain on the mill; it also lets the material sit as a lump at the bottom and protect itself from the chain. So feed it slowly, and stop and empty it before too much gets in it. How big a batch you can run depends on the dimensions of the components you scrounge for this project, and you'll quickly learn after a batch or two how much that is.

The chain should be replaced as it wears. If you use this little mill a lot, you will wear out the brake drum too, so keep an eye on it. When a spot starts to wear thin, weld on a layer of hard-surfacing rod. You can expect to do this fairly often if you end up grinding much rock. Coal, charcoal, and Sheetrock don't wear a mill much, but most ores and magnetite sand wear steel quickly.

Mini Miners, Throw Off Your Chains

The next level of mini impact mill is very similar to the chain mill, but instead of a chain flail it breaks the material with rigid radial hammer arms mounted on the shaft. The advantage is that these arms also act as radial-fan blades, which means that they create a vacuum at the center and high pressure at the periphery. Because of this, it creates a positive feed at the center and discharge pressure at the outside of the drum, which lends itself to a continuous-feed configuration.

The flow of air created by the radial-fan effect of the hammer arms can be used to both carry the material through the mill and to classify heavy material as to size. By varying the height of the discharge pipe, you can select the fineness you want to leave the mill: the higher the discharge pipe, the finer the ore must be milled before the air will carry it out. Charcoal, on the other hand, should be contained in the mill until it is milled, as fairly large chunks will ride out on even a slow stream of air.

An important consideration in building this second type of impact mill is that the rotor must be very well balanced if you want the bearings to last. No matter what type of rotor you make, the balancing procedure is the same: spin it by hand a dozen times or so, and each time mark the arm that is down when it stops spinning. Add weight on the opposite side. When using the machine, check it for balance every now and then, as steel striking rock wears quickly but not always evenly, so a mill that is balanced when new probably won't stay that way. If you weld on hard-surfacing rod as the hammer parts of the rotor wear, be sure to always apply the same amount to each opposing arm.

Rotor Types

The two easiest rotors to build are the straight-arm and

the disc type. The straight-arm rotor is simply a steel bar attached in its center to the shaft with a lock collar. Mounted at each end of this bar are chunks of steel to act as hammers. The disc type is a flat circle of 1/4-inch steel plate, slightly smaller than the inside of the brake drum, welded to a lock collar, which secures the disc to the end of the shaft. Bolted or welded radially to the face of this disc are lengths of angle iron, which in turn have replaceable flat iron wear faces bolted to them. The advantage of the disc-type rotor is that it acts better as a radial fan to help move material through the mill and tends to keep abrasive material from coming out the shaft side of the mill and ruining your bearings.

ROLLING ALONG

Another type of mill that can be improvised is the roll mill. A good set of rolls can reduce ore to a fineness that will release most sulfides and also suffice for our other milling projects at hand. One advantage of a roll mill in a survival situation is that, with a flywheel, you can rig it with a hand crank. We include it here to provide you with another option. After all, in the final analysis, survival is often a matter of having provided yourself in advance with options.

A roll mill is not unlike a heavy-duty clothes wringer, made from steel and with the rollers mounted horizontally opposed. The size of the discharge is determined by how close the two rollers are: for a powder, they are mounted almost touching. The larger the diameter the wheels, the larger the material you can feed. One wheel is usually an idler, with the other being power driven. The rolling action feeds the material into the rolls, where it is crushed and drops out the bottom. If you make the rolls adjustable, you can open them up to feed larger material, then close them back up and run the same material again to achieve a greater fineness.

For improvised rolls, the handiest off-the-shelf thing I've found is the cast-steel doughnut weights that are hung from a fire door to ensure its closing. They're usually about 6 inches in diameter and 3 or 4 inches thick, with a hole already through the center. You can also use a short section of heavy, large-diameter pipe that has been filled with concrete around a shaft, if you hard-surface the outside of the pipe. If you rig it with a large flywheel, you can hand-crank this machine while a partner slowly feeds ore into it.

Roll mills, like jaw crushers, will sometimes spit material back out. When you have assembled a set of improvised rolls, finish the job by installing cheek pieces on either end of the rolls to help contain flying pieces. And *always* feed rolls by rolling the material down a chute; if your shirt cuff or finger gets between the rolls, you'll be up to your elbow before you can react. A forearm that is 9 feet long and a quarter-inch thick will not support a musket.

CONCENTRATE ON WHAT YOU'RE DOING

Now that you have the ability to break rocks down to a size that releases what you want, the next thing is to separate out what you want. It may be sulfide minerals, or it may be magnetite sand. They both are heavier than their host rocks, so both will concentrate satisfactorily by gravity methods. Sulfides can also be separated by oil flotation as we mentioned before, and magnetite can be separated magnetically. We'll detail those methods a little later.

Hand Panning

The simplest form of gravity concentration is the old prospector's hand pan. Fill the pan roughly a third full, add enough water to make a soupy mix, mix it up by hand, and shake it violently back and forth to settle the heavy material. Then tilt the pan forward a little and use a rotary motion, tipping the pan a little more forward with each rotation until you start to spill water and wash off the lighter material. Add water as you need it. Keep going until you have only a crescent of what you want—sulfides or magnetite—left in the bottom of the tilted pan.

Rocking Sluice

The most productive simple gravity concentrator is the rocking sluice. It is sort of a cross between a gold pan and a conventional sluice box. Its advantage over a conventional sluice is that it uses much less water, and over a gold pan that it is much faster. To operate, mix the material with water and dump it at the head of the sluice while you gently rock it like a cradle. The heavy material will be caught behind the riffles, just as in a regular sluice.

Dry Washers

The heavier minerals that you want to stock your aboriginal lab can also be separated by gravity in a *dry* environment. If you're patient enough you can do a fairly good job dry panning, which is pretty much the same as wet panning, but using a more energetic shaking motion and shorter strokes. And a lot more of them—dry panning is slow work.

The most common dry mineral concentrator is the air or bellows-type dry washer. It is arranged very much like a conventional sluice, with fairly conventional riffles, which are set at a steeper angle. The sluice frame itself is set at a much steeper angle than a wet sluice box, and the bottom of the dry washer is made from fine-grained cloth or screen.

Below and integral with the "sluice" part of the dry washer is a large bellows. When this bellows is pumped, air puffs up through the porous cloth bottom of the "sluice" and blows the lighter material on down, leaving the heavy concentrates in place behind the riffles. Some dry washers also apply a shaking motion in concert with the puffs of air. They can be hand operated or motorized. They are very dusty to operate, so wear a mask.

Some dry washers use only a violent shaking motion without an air flow. These are called dry shakers. The principle is the same as dry panning, and the construction is the same as a conventional wet sluice, except you use vertical instead of slanted riffles and a slightly steeper angle on the box. The "sluice box" is suspended from four cot springs under slight tension, and an orbital sander is firmly mounted on the center underside. Milled material is slowly and evenly fed behind the top riffle. The orbital action of the sander shakes the heavies down where they are trapped by the riffles and shakes the gangue on over the rifles and discharges it to the tailings. A dry shaker is still dusty to operate, but not such a nuisance as a bellows-type dry washer.

We should also note that there are innumerable tools, devices, gimcracks, and black boxes being promoted for weekend prospectors designed to capture the last speck of microscopic gold. Some work and some don't. They're all expensive. For our purposes here, the simple tried-and-true, old-timey machines will serve as good as anything.

Flotation Separation

We mentioned earlier the phenomenon of sulfides forming a floatable froth when mixed with oil, water, and air. Some minimalist flotation cells would be the concrete mixer or shaken coffee can suggested. But the simplest production device to mix ore, oil, and air would be a Pachuca tank.

The Pachuca tank was devised to mix gold ore that was being leached with cyanide. Cyanidation of ore requires air in the mixture, and it requires agitation. The Pachuca tank provides both, and does it with no moving parts.

The Pachuca tank is simply a vertical tank, at least three times as high as it is wide, with a cone-shaped bottom and a vertical inner tube or pipe roughly one-third the diameter of the tank. Air is fed at the bottom of the cone and rises up the inner pipe. As it rises, it expands, and as it expands it increases its upward velocity. It sucks up material from the bottom of the cone and carries it up through the inner pipe, mixing it with oil and air as it does. Material comes out the top of the inner pipe and falls back to the bottom to repeat the cycle endlessly. This efficient mixing apparatus was not designed as a flotation cell, but it works very well as one, and it is very simply made. The taller it is, the more efficient it is, but a relatively short one will suffice for our purposes.You can make one from large plastic sewer pipe or an old hot-water tank—use what you have. The cone-shaped bottom is important because it funnels the material back down in front of the air stream so it can be blown up the inner tube again. If you don't have a tank with a cone-shaped bottom, you can possibly adapt a large funnel. The easiest method is to simply use a good, rich mix of concrete or mortar to build in a cone-shaped bottom. A hand skimmer (a colander screen with a fairly tight cloth over it) will do to gather the sulfide-collecting foam at the top of the tank.

If you do not have an air compressor, many shop vacuums will provide enough pressure at their discharge to operate a small Pachuca tank. If you do not have an air source, you can rig to a vehicle exhaust. The pressure required is not great, and in this instance we are using the "air" flow to provide agitation and not for its oxygen content.

Many lashups can be improvised as flotation cells, as long as they provide a way to mix oil and air and water together. Feel free to experiment.

A Magnetic but Amicable Separation

Magnetite sand gets its name from the fact that it is magnetic. Magnetite ore that had been naturally magnetized by lightening was called lodestone and was used as a magnet in ancient times. The ability of magnetite to respond to a magnet makes separating it a simple matter.

The simplest device is a magnet in a plastic bag. The next simplest and much more efficient device is a magnet on the end of a short rod placed inside an aluminum salt shaker. Small prospectors and miners use it to remove the black sands from placer gold concentrates; we'll use it to remove the black sands we want to keep. In use, the magnet is lowered to the bottom of the salt shaker and immersed in the concentrates, then lifted up with a load of magnetite. The plunger holding the magnet is then raised and the magnetite falls away from the shaker into your container. Then you go after another load. Like using chopsticks, once you get the hang of it, you will be amazed at how much material you can move.

And Good Luck!

Now, during the process of gathering, milling, and concentrating ores for sulfides and magnetite, odds are fair you may end up with some gold if you're doing this anywhere in the west. Is the gold worth saving? Sure, all gold is worth saving. Is it worth messing with? That depends on how much of it there is. Is it easy? Quite easy, compared to mining and breaking the rocks to begin with. How do you do it? It's really fairly simple, but *that's* a story for next time.